ADVENTURE
into
SUBSTANCE
(big bang indeed)

A Revelation

received by

Paul Wiegand B. Arch.

concerning

The beginning and Purpose of all That is.

Inner Garden Publishing Co.

Paul Wiegand
30 Victoria Ave.
Belleville, Ontario
K8N 1Z4

ISBN 0-9681194-0-9

Contents

Introduction

"Adventure into Substance" is an allegorical poem concerning the beginning and purpose of all that is. The ideas and concepts in this poem came to me as a "knowingness" during an experience of great clarity of understanding. This occurred recently when I was awakened early in the morning. For several hours many thoughts concerning the universe, God, creation, and the purposes of man on earth kept tumbling about in my mind. Some new ideas I had never thought of before came to me. As I meditated on these things a picture of how it all happened formed in my mind and soon became very clear. This experience of clarity was repeated for several days in a row. Each time I was awakened early in the morning. After several days of this it came to me through the same "knowingness" that I was to write down the things that had been forming in my mind.

Several days later I began writing what I had come to "know". I attempted to write in ordinary descriptive prose. As I wrote the ideas were developing as an allegory. They also seemed to be coming out of me in a rhythmical poetic style that I would call blank verse. When I yielded to the impulse to write in this manner it began to come with surprising fluency.

I had never before written in this form or style so I was quite surprised when I read back what was written. I would call this phenomena a combination of the word of knowledge and inspired writing. It was not channelling or prophetic utterance or automatic writing. Along with the fluency there was much of the writing that required considerable effort, revisions and editing.

It was a puzzle to me as to why I was led to write in allegory and in verse. Allegory is a story that reveals truth. The story may not even be a factual account. My story of God, Lucifer, Adam and Eve, to my way of thinking is not a true factual story. But I do believe that this story reveals important truths that need to be considered at this time. The ideas are quite revolutionary and may be more palatable in allegorical form. If you don't agree with the new ideas you can say "well its just a story" and it may be less offensive than a theological pronouncement. In religion allegory has long been a way of presenting truth. It adds plot and interest to what might otherwise be forbidding theology.

The allegorical way of talking about God in human-like situations may reveal things about God better than doctrinal statements. Our story talks about God, for instance, as being bored and lonely, hardly the proper way to describe a transcendant Deity but possibly revealing another side of God that we haven't thought about. Anthropomorphism is attributing human characteristics to God. Our poem uses this figure of speech frequently. While I consider God "beyond personality" I make no apologies for speaking of Him in a personal way. Because I am a person, He can be no less.

The poetic nature of this work I believe is for the purpose of lifting God out of theology into poetry. His poetic and artistic nature is exemplified in the universe and man His creative masterpiece.

I have loosely followed the Biblical account of creation but many times taking a totally opposite direction than the traditional interpretation of the Bible.

For example you will discover in this work that God's attitude toward the "tree of knowledge of good and evil" is quite the opposite to the usual understanding. In our account God does not forbid Adam to eat of this tree but says that it is necessary for Adam eventually to eat to know both good and evil so that he can make an informed decision as to what he will chose.

There are many other departures from the Biblical account. In fact the writing suggest that the Bible account is actually part of a conspiracy to hide the truth from men. In spite of this I believe there is an abundance of hidden truth in scripture. The scripture references throughout the writing are not for the purpose of quoting authority, but rather references to passages where hidden meaning has been seen.

This writing is all about matter and spirit. Such relationship in its totality I call "God" although I would much prefer to use the term "Mother-Father Source". Our English word for matter is derived from the Latin "mater" meaning mother. I see the spirit aspect of the Source as Father. But in consideration for practical and poetic expediency I use the term God instead of "Mother-Father Source". But I am not referring to the God of religion made in the image of man.

Also for expediency I have used the pronoun "He" when referring to God. If there were a word in the English language to describe one who is both "him" and "her" I would have used it. For I believe God is both male and female. But there is no pronoun to refer to such an androgenous entity. I could use "it" but consider that too impersonal.

Feel free to substitute She or It for He as you see fit.

Adventure into Substance

Something cannot come from naught.
So everything that is
both of matter, and of consciousness,
has always been.

Matter exists and has always existed.
Consciousness exists
and ever was aware of its existence
and the existence of matter.

This totality of consciousness and matter
was aware that it was everything
And there was nothing else outside itself.

Aware
of what it consisted, and all that it was.
Aware that it was spirit and matter,
singular and plural,
male and female,
light and dark,
hot and cold,
good and evil
all together and therefore somewhat neutralized.

Conscious and aware,
and if it were possible it would be,
me thinks,
quite bored in its neutrality.

It was everything
but wanted to be something
and thought about becoming.
And thought to speak a word
to change and re-arrange itself.

But never more or less
than ever was
or ever it will be.

We could call this conscious matter,
and if consciousness could think,
thinking, conscious matter.

And from this glorious, thinking, conscious matter
came the Word.
the cause of all effect.

This Word has been called God
And was in the beginning with God (John 1:2)
And through this Word
all things came to be from all that was.

And from this thinking came all thought.
And from this knowing everything
that ever could be known.
And from this being
everything
that ever there could be.

✦✦✦

Now God saw that everything was formless
and darkness mixed with light,
everywhere.
And thought
this dark-light could be made to be
of greater worth
if separated from itself.

And so God spoke the Word
and said
"Let there be light" (Gen.1:3)
and separated the light from dark (Gen. 1:4)
within Himself.

And lo
beautiful brilliant light
along with and beside
the absence of light called darkness.

Suddenly form and shape appear,
shades and shadows,
dimension and depth and space

And hand in hand with space came time.
For one without the other cannot be.
Within the vastness of eternity.

✦✦✦

As light and dark
so good and bad or love and hate.

God knew that He could not become
what He desired to be
unless there was a further breach
of His neutrality.

What wondrous thing this love
He saw
amidst the whole of all that was.
What wondrous qualities,
priceless treasure,
singular delight.

He thought
of all the things there are that He would want to be
He would choose love.

But knew
that this He could not be
except love separate from all that is not love
and from the whole of what He was
since all eternity.

He looked
and saw this residue,
the darker side of what He was
when all that was not love be separate.

What hideous array of evil hate and madness
anger, fear, and murderous intent,
and qualities one would not seek
except to know their opposite.

And in His singularity
He could not see the harm
to separate the two.
Love would prevail and overcome
He knew.
It had the edge.
As darkness over light could not prevail (John 1:5)
So love o'er hate must triumph in the end.

And so the Word of God went forth again
"Love be separate from all else."
And it was so.

Then He became what He desired to be.

And God was love.

As is He
even to this day.

✦✦✦

Now God was aware
of the great potential of His thinking.

So He thought adventurously
and revelled in the vastness of His mind.
He thought imaginatively
in His boundless creativeness.

His thoughts were limitless in scope,
magnificent in scale,
and intricately focused in detail.

In His mind He saw;
galaxies and stars
and whirling nebula
suns, and moons and beauteous worlds in space
protons, neutrons, atoms and cells.
and endless fantasies.

God revelled in His adventurous mind
but knew
that thinking would not make it so.

And He would have it be.

So to make His thoughts become reality
this matter must be separate,
He thought,
from all that spirit was.
and then be separate from itself.

Such fusion could only come to be
with vast amounts of energy.
The only source in quantity
to do the job
was love.

So He must speak again to make it so.

And God said
"in love I decree
that it shall be".

And it was so.

✦✦✦

Then began
a chain reaction
precipitated by this love
which grew and grew
accelerating wildly
until it reached the speed of light.

And then
a joyous explosion
such as the vastness of time and eternity
had never known.

The energy of love,

A blinding light and deafening sound,
the power of motion through reaches of space.

And in the wake of this cosmic bang
clouds of gaseous matter
and swirling tendrils of primeval mist
Birthed stars
and clusters of stars
and galaxies spiralling through space.

Big bang indeed.
God knew
one day
that special breed called scientists
would find a formula
to calculate
the energy within this "matter".

But could they ever know or even estimate
the Power of Its love?

God gave His substance to become
a billion billion
or more (God only knows)
sub-atomic particles
the building blocks of this material universe

To form
in the ages to come
and manifest
the glorious Creation of the vastness of His mind

Before the foundation of the world.

And so each particle contained
the nature that He was
and more
the energy of love,

The substance of God dispersed
in cosmic proliferation.

And still the mind of God
not subject to the laws of matter
remains in-tact
alive and well and moving everywhere
even to this day.

Now God is all
in the glory of His mind

And in all
in the substance of His universe. (Eph. 1:23)

This is the inheritance of God
as spoken thus in ancient time
to Abraham of old,

"Your seed
shall be
as sand upon the sea
sons beyond number
as the Stars of the sky
that cannot be counted". (Gen. 15:5)

✦✦✦

And once again God thought
of who and where He was,
of His magnificent singularity.

His body the substance of the cosmos

His mind the consciousness of spirit.

And God said
"I am that I am" (Ex. 3:14)
there is no other.

And God was alone
except for His thoughts,
and felt His solitude.

Then He considered what His thoughts could do.
within the vastness of His mind

He could create a being like himself
within His consciousness
with whom to share His thoughts.

Such a product of His own thinking
would choose as He decrees.
No problem here.

Though Love's separation exposed the bad
such a being of His mind
would ever choose the good
for in himself He willed it so.

So He created in His mind
a being of Light
in glorious splendor
like Himself.

Now God was not alone.

This entity though only in His mind
seemed as alive as He

And He talked to this fantasy of His mind
as we've been known to do.

He shared with it His consciousness
and gave it ability to think as He

to ponder and create.

and they had fellowship, the two.

Then
made within their minds
more like themselves,
and soon
a vast array of entities of thought,
each reflecting aspects of what God was.

To each, He imparted gifts and attributes
so none alike
in the diversity of His great mind

Each with consciousness to think as He.

And you
were a thought in the mind of God
perfect in pristine beauty.

And I was there
unspoiled
fulfilled.

And we but two of a cloud of souls
formed and chosen by Him.
Created blameless and in love (Eph. 1:4)
to inherit and enjoy
the magnificent kingdom of His mind
before the foundation of the world

All living spirits in the spiritness of God.

To His created ones He gave names
and appointed tasks
some caretakers of His worlds,
some to explore the great expanse of space,
some to create as He,
but all to enjoy as He enjoyed
the vast profusion of His mind.

✦✦✦

The shining creature he fashioned first
He called Lucifer.

And he it was who spoke for all the rest.

To Lucifer God said
"You are my pride and joy
how wonderful it is
to share my mind
with you and all your kind".

Lucifer acknowledged his Creator
and said to God;
"You have created us in your glorious imagination.
You have endowed us with ability to think as you.
We are mind of your mind
and we enjoy the spirit world of your spirit mind.
Your imagination seems to know no bounds.

Yet

We do only what you decree.
we are not free
as you, to be
what we ourselves decide.

And God said;
"Lucifer 'tis true
Spirits knows but the limits of my mind
but that is limitless.
What other freedom is
that could ever be desired".

And Lucifer replied;
"Indeed
We can do anything you will us to,
fly to the limits of your creation,
and enjoy
the boundless pleasure that your thoughts afford.

But what is this "matter"
the real substance of which you consist?

We are spirit as you are
but of your substance we know nothing.
What mysteries lie within your form?
What pleasure
unknown to the spirit world
lie hidden within
the density of your singularity?

We have bodies, true,
but they have no reality
as yours.
Our bodies are but phantoms of your fantastic mind.
Could we not know the nature of your material substance
and still
the freedom of your mind?

Could we not become as you?"

The words of Lucifer
brought forth a stirring
in the material substance of which God was.

He knew Himself
and the nature of His being
of matter
and its energy of infinite potential.

God thought

How gloriously superior would be
this wondrous creation
if endowed with the substance of which I am

And free to be
themselves

And Gods in their own right.

How much grander would be
such expression of my Love
than the thought games
I have so long enjoyed
and loved to play

With the creation of my mind
through all these eons of time".

What if
these thoughts of mine
who share my consciousness

Would share my life
and be as me

And free
to do as they decide
and be what they decide to be.

And if they had a body of their own
which joined their minds to be like God
in every way,

What if they had the right to choose
and the ability

Would they choose right,
light over dark,
good over bad,
reject the residue
for love above all else?

And if they did
this love would surely bind them close to me
and to each other.

But if perchance they chose
one fragment of the residue in place of Love.
what chain reaction might ensue
to drive us far apart.

You can be sure
the mind of God considered all the possibilities

And knew

That some would choose the wrong.

Yet He could see that they would have the chance
to choose again
and yet again.

And each choice for the good
would make the next one easier.

And in the end
love will overcome.

God saw the nature of reality
and knew the end from the beginning (Isa. 46:10)
that good would finally prevail,

Although,

When faced with freedoms probabilities
He could not tell how long.

But His decision would be based on love
not fear.

Then He became acutely aware
of a surge of energy welling up within Himself

The energy of love.

And God so loved the world (John 3:16)
and His sons
and the whole created universe of His mind

That He gave of Himself,
His right to choose for them.

So He must separate Himself again.

His thoughts from Himself
their thoughts from His
their will from God's.

And as a parent eventually
will put His children
on their own

So God

The vast creation of His mind
sets free.

And God said

"I set these creatures of my mind
free
to be themselves.

That they may know
the joy and beauty of reality
and be as me."

And so was launched
the great and wondrous experiment of the ages.

In substance God gave Himself
to form the universe.

And now
in spirit He separates again

No longer one.

And this, the Lamb of God
slain from the foundation of the world (Rev. 13:8)

Divided
as the sacrificial Lamb
upon the altars floor of time
with smoking pot and blazing torch
between the pieces (Gen. 15:17)

To show to all
the mystery of sacrifice, and God, and love.

✦✦✦

Then God searched the vast recesses of His mind
to find a way
to have this great creation of His mind,

Free spirits now,
for each to become as real as He,
in matter and in substance
as was His glorious
singularity.

He thought
I must prepare a place (John 14:2)
within the cosmic universe
for them to call their home.

In His mind He saw that special place

Prolific in resource in beauty and variety
a place His sons could know His joy
and boundless happiness.

He spoke the Word
to bring it forth
from out the wondrous galaxies.

And there in time and space
out of that whirling mass of suns and stars
and cloudy gaseous matter
we call the milky way

A solar system grew.

And the spirits of Gods mind were there.
as witness to this glad event
"And the sons of God shouted for joy". (Job 38:7)

And one special jewel called "earth"
sped in silence around its sun
a billion years in preparation

Whose molten core broke through its barren crust
erupting floods of lava pouring out
releasing gaseous vapour
long since trapped within
formed pools in hollows of its naked rock.

Then oceans,
deepening,
covering all,

Until the lava cooled

And land appeared

Then majestic mountains looming high
stood in silent splendour
as if to wait and watch with God
the primal pool of water down below

In wondrous expectation.

And then another billion years (God only knows)
the water swarmed with tiny bits of life
evolving slowly
until the oceans teemed with myriad of kinds.

While naked rock was carpeted
with many shades of green
and witnessed an invasion from the sea
a million creeping, crawling things.

And birds to streak through skies of azure blue.
And presently a carnival of creatures great and small
to play upon the land.

In forests deep, lush meadows, and sandy dunes
beside its lakes and rushing rivers,
quiet streams
grew herbs and trees with fruits profuse
in numbers and in great variety.

Then

Upright creatures walked upon the earth.

And spirits watched with bated breath
until the time was ripe
to see
the mystery
long promised to the sons of God.

✦✦✦

Who will go first?
and who must wait?
and who may choose to stay
within the safer spirit realm?

God wondered, then said

I will let each choose for Himself
the time and place to incarnate

✦✦✦

Then a company of spirits came

To volunteer

To test the waters of reality,

Spy out the land, and see
if spirits could indeed
take on a garment of substance
to become like their creator.
in mind and matter both.

Of mind they knew a lot
but matter was a part of their creator
they knew naught.

Nor did they know,
without Creator's guidance
according to His Will,

How to be free

Within this world of substance.

Then Lucifer came forth
and said;

"I've been the leader here in spirit realms.
should I not be the one to lead the way
into this unknown realm
of substance and free will,

I, first born of the spirit world,
who knows and loves his maker well,
and of the mind of God
familiarity".

To this God did agree.

✦✦✦

And to prepare the way for them
God planted a garden (Gen. 2:8)
wondrously profuse.

With trees of all varieties
for beauty and for use.

And a tree of life (Gen. 8:9)
Enabling them to live forever if they choose
upon the earth
or anywhere within the universe.

And a tree
that has been called
"The knowledge of good and evil".

Then to Lucifer
and his company of volunteers
God spoke these words of preparation.

"You shall be free (Gen. 2:16)
to eat of every tree
I've planted there for you.

There is no law to say
this is right or this is wrong

But if you would be
as God
upon the earth

You shall eventually
partake of that one tree called
"The knowledge of good and evil" (Gen. 2:17)

Then to be - love, as I
you must separate love from all that is not love
and as God
choose only Love
and leave the residue.

And if perchance you choose the wrong
yet will you have the chance to choose again

And yet again

Until you see
the wisdom of this love."

There also is a tree
that you may eat
to live forever on the earth
or anywhere within the universe,
The tree of life.

And if you so choose
you may return to spirit world.

Just lay down and fall asleep and die.

Quite simple, no big deal.
others will return your body to the earth.

And in an instant
you will be
with me
and others you have left behind
within the spirit world.

Here to plan your strategy.

To stay
or to return as case may be.

A body I have carefully prepared for you (Heb. 10:5)
through many years of evolution

Homo Sapiens called man

And into it I will breath
your spirit
that I created in my mind
with which I fellowshipped so long
in blissful realms of thought

And as a living soul incarnate

Still

I will be there
to walk and talk with you
within the garden that I made.

And if you stray
and fail to hear my voice

Remember

It is true I will never leave
or forsake
you. (Heb. 13:5)

Now it is time to go".

Nor was the journey far
'twas but a breath away.

Into the man He created
God breathed
the breath of life

And man became a living soul. (Gen. 2:7)

First Lucifer
light being of the spirit realm
with new identity

Becomes a man
or "Adam" if you will.

Then soon to follow all the company
of volunteers.

✦✦✦

What Joy

What a surprise

To be alive

A spirit still,
yet in a body of flesh and bone and blood.
and conscious of a world of matter all around.

A world of sight and sound
to touch, and smell, to gaze upon

Living matter of energy and love.

Beautiful
beyond all expectation.

But still the one He loved so well was there

And spoke to Adam;

"I have given you every herb bearing seed
every tree and plant and fruit for food,
and flowers,
beasts of the field and forest, birds of the air,
and every creeping thing upon the earth
wherein their is life;

To name, and tend, preserve and to enjoy" (Gen. 2:28-30)

And God and Adam saw that it was good.

✦✦✦

Then God remembered a principle

That good could not have been
except it separate from bad

As light from darkness
spirit from matter
and love from all that was not love.

So now He also saw
how much better it could be
for man upon the earth
to separate once more
male from female.

So Adam fell asleep (Gen. 2:21)

And when he woke
he could not tell how long he slept
it seemed no time at all.

But there before his wondering eyes
he saw
standing gracefully
upon a grassy knoll
the partner God had promised him

A being from the spirit world indeed

Within a body
whence he could not tell
beautiful beyond compare

But not like his
the difference most agreeable.

"Welcome Fair one
into a world of wondrous mystery

You shall be my pride and joy

And I will love you
with the love of God
so freely given to me".

✦✦✦

And so they walked off
hand in hand
As God's upon the earth
in wondrous singularity

To claim the world prepared in love
for them
and all the spirit realm
who chose to take upon themselves
the garment of reality.

Which Adam and Eve now enjoyed,
partook of its pleasures,
and ate of every tree
that they could see

Except so far.
the tree of life

And the tree
of knowledge of good and evil

Although they knew
as God had said

Eventually

They must partake of that one tree
so they could see all that there is to see

And still choose love above all else.
to be like God
in love and free.

✦✦✦

So one day by chance
they came upon that tree
and wondered what to do.

Adam looked around the company
of Incarnate spirits he knew well
gathered there to see
what he would do.

And he their leader
knew
that he must show the way.

Was it not He who said to God
"we would be
like you"

So far he knew what it was like
to be in form and free

But did not know
as God, both good and ill

Though he knew love
for he knew God

But still

Evil was an unknown mystery.

So he did hesitate.

But Eve
whom God had given him
stood boldly forth, plucked the fruit, and ate
and gave to Adam. (Gen. 3:6)

Then they did see
the love of God more clearly than e're before.

Against all that was not love
a fearful sight.

And fear it was that done them in.

Those who watched
saw Adam's face
(though on the face of Eve not quite so much)
a look
unlike anything they had ever seen before.

None knew the look of fear
for they as yet had not partaken of the tree.

But following their leader they did eat.

And saw what Adam saw, and more
some avarice and pride, some selfishness and greed,
some deceit,

And thought in frantic fear
to spread the word around
that God forbade they eat of that one tree.

And if they did, dire consequences would ensue
that they would die (as if to die was bad)

It was a lie of course

But those who listened to their tale
dug out of the hideous residue
a thing called "guilt".

So guilt and fear ran rampant

Through Paradise

Where God intended naught but love.

✦✦✦

Some took it on themselves
to try deceit on Adam and Eve

And spoke
as if to imitate the voice of God.

As in the cool of day
Adam walked with Eve.

"Where art thou Adam?", they called (Gen. 3:9)

Then in both, fear took hold

They hid themselves
as if their nakedness
could displease
the one who made them so.

And so in foolish self defence

Adam blamed Eve

And Eve the serpent

But in the saddened heart of God there was
no blame at all

Nor even to this day

✦✦✦

Soon conspiracy arose
amongst the company
of those
who ate

They tried to use their influence

And sought
to spread the word of fear of God
with talk of vengeance, jealousy and wrath

Which God was not.

But thus they did conspire
to drive man from the garden
with curses vile.

"Go from this place and sweat and toil
in sorrow all your days (Gen. 3:17)
lest you eat of this tree of Life and live forever"

It was not God
who cursed the land
Or Adam, or Eve, or serpent there.

Not flaming sword
but fear
and guilt
kept man from eating of that tree
all these years

To ever walk as God
upon the earth as it was meant to be.

And this conspiracy
holds sway

Even to this day.

✦✦✦

And there were those
who chose

To exercise power o'er all the rest.

And saw themselves as gods
(although they were, but not of fear and guilt)

Assayed to rule the earth
in fear.

Most of the others fell in line.

"This residue that was not love
must be controlled" they said.

"Its not to choose but regulate".

And so arose

Laws and rules, punishment and controls,
to make men do the right

Then
put in man the fear of death

And punishment

And Hell.

✦✦✦

But worst of all
they tried
to hide
the voice of God.

As Adam was deceived so many more.

"Follow our decrees and rules and manuscripts
and books galore
to know what God would say to you."
"Only a few select ones hear God speak", they said
"It is a rare phenomena".

So many have not listened for His voice.

And do not know

Who they are

Or why they're here

And they have been within the bonds of this conspiracy
for countless generations

far too long.

✦✦✦

And yet another group
of those who came to spy the land
and saw the dangers lurking there

Said

"It is indeed a land of milk and honey

But there are giants
far too big for us to overcome

Are you sure
that we were meant to stay?"

They say
"'Tis but a school to learn
and then return
to where we're meant to be"

A school indeed -
to learn to live in matter's form.
But what the lesson here
to learn to live in the land of bliss
from whence they came where choice is not an issue,
or freedom not the goal, matter
but a mystery known to God alone.

Having seen the darker side of God
they say it is too hard
to choose the right from wrong.
And freedom much too difficult to handle

And matter too inflexible

Not like the spirit world
where naught save thinking is required.

Matter is but illusion they say

And Life not real.

Surely would it not be
The best solution

To return to spirit world
where all is safe.

So let us fly away
they say
and be forever with the lord

Not knowing He is here to stay
with us
where we are meant to be.

But in reality

We are He

Mind of His mind and
substance of His substance.

But many on earth
And even in the spirit world
who saw it thus

Conspired to promote
the safe return of all to God where they belong.

They say
to save them from themselves.

And many in this world see it thus

'till they return

Where eyes are opened

Then cannot wait to come again
and scheme to try to get this time
much closer to their goal

And so they do.

✦✦✦

Of all of this
there is much we do not know
or understand.

But this we know
the promise made to give the Sons of God
the chance to be like Him

God honoured
and honours still

For you are one of those He formed
within the confines of His mind long years ago

And I am here
to taste the substance of Himself
and grow to be one day like Him

So we can walk as Gods upon the earth
in spirit and in form
as He who walked in conscious singularity
before the world began.

Nor is the process easy
we have learned,
as was it in the spirit realm
to do the will of God.
For there we had no choice
save to be the great perfection of His mind.

But in the flesh perfection comes with effort
and through a process
diligent attention,
sincerity of will,
and many trials,
and often errors and mistakes.

And if we fail and falter
yet are there more chances
to return and try again.

For this material world is meant for us to stay

Or if we choose

To come and go at will.

But none the less to know
that "matter"
is but particles of God.

To which we join our souls
in a mysterious relationship of love.

To live in this material world
and have the mind of God
and love our brothers as we love ourselves
and to be as He created us to be
before the world began.

And yet like Him in substance
and in mystic union of body and of soul
mind and matter
in consciousness of all that is.

"Matter"
so strange to we who only knew
a spirit world

We seem not to know
that God is in it, every stone and every tree
and every atom, every cell
hidden away
to be revealed
in the last day

In countless sub-atomic particles

In substance

Energy

In power and light and love

And in our predilection
oft we count this "matter" less than spirit
dismiss our form
to fly away to spirit realms above
we think superior to this lowly earth

Rejecting matter
as the stone the builders set aside
destined to be
the chief and corner stone (Math. 21:42)
of our reality.

And in our struggle to adjust to freedom and
to "matter's" form
we have a Helper

Holy Spirit He

None other than the mind of God
Whose mind remains in-tact

Though substance scattered far and wide
and often hidden from our view.

His mind and voice are there for us
if we but listen to its sound
for us to know the Father's will
and plans
and purposes.

Yet is the outcome certain.

And in the mind of Him,
Who saw it all before the world began
we were created perfect, He could do no less

And our perfection is assured
predestined in the mind of God before the world began

We wanted freedom
And in love receive the same

To be as He.

So freedom is our goal.

But freedom means to choose
and choose we must
though sometimes short of His perfection

Though we can choose again
and many times

'Till our desire to be like Him
and His desire to see perfection
coincide

✦✦✦

So this is what its all about
for spirits of the mind of God
of which we are

To incarnate
and live on earth, joining spirit to matter

And in freedom choose love above all else.

So we can walk as Gods upon the earth,
and be like Him in every way.

And stay, or go at will.

This has been the process
through ages and millennia

Since Adam

We have tried yet often failed
and tried again, and learning all the while
amidst the darkness,
confusion and illusion of conspiracies
it has been hard to find our way.

Earth's education has taken time and toll.

And all the while
God's hand has guided
within the restraints of freedom's choice.

But God knows
and it was prophesied by seers of every age

That we await a day to come

When we will see
the consummation of His plan.
to make His thoughts as real as He.

When right will triumph over wrong
and men will awaken
to who they are
and what their purpose here.

God did not know how long
"Tis true
His sons, the product of His mind
to be like Him in every way
When faced with freedom's choice

But knew

Eventually it will be so.

And we
like He
in patience wait the day
to see the kingdom of this world
become the kingdom of His mind
which He foresaw
and knew
that it will come

Though seems it long to wait
to see it fully manifest

Through times of darkness and despair
there have been lights in every age
appearing on the world's stage

To give us hope

But ever waiting for an age to come.

And in this day
there seems to be
new rays of hope

As many eyes are opened

Some have discerned the probability
and say
that we will see it in our day.

Is it possible that we have come
to the end
of this long, long journey?

Could it be
our eyes will gaze upon
as Adam gazed in wonder that first day

A paradise restored most beautiful in form

Beneath a panoply of love to cover all.

Who knows what wonder lies in store
and glories yet to come

Only the greatness of God's mind
could articulate
the golden age of God upon the earth.

We do not know

Although

It would appear
That we are close enough perchance
that we could turn the tide.

A conspiracy it was
that set us off our course

Could we not launch one of our own?

To spread the word of who we are
and what our purpose here

That God is not the vengeful one
that they made out.

Perchance we could haste the day
that men will see

It was a glorious conspiracy

Of God

To give us life and love and freedom and reality

All out of Himself

Who is and was eternally.

✦✦✦

When we become
in substance and in mind as He

Then we in turn
can give ourselves
to bring our thoughts to life

And in the doing be
Not less but more

As He
from singularity
became a never ending expression of His Love

And see
the glorious expansion of ourselves

In love
throughout eternity.

✦✦✦ The End ✦✦✦

Post Script

In my introduction to this volume I said that an allegory is a story that reveals truth. I have told my story and the truth may or may not be self evident. But for any who wish to pursue these ideas further or seek further clarification I will add a few notes and observations concerning the things contained in the preceding poem that I saw so clearly in those early morning experiences.

In this post script I will express these things in prose as I tried to do before I was led to write in verse.

My only authority for all that is written herein is the "knowingness" within myself. I understand that while this may be truth to me it will not necessarily be truth to you unless the Spirit of Truth gives you your own "knowingness" concerning these things.

Matter and Spirit

The poem talks a great deal about two aspects of

reality, matter and spirit. Matter has been the subject of scientific investigation for many years. They tell us it is made of up of molecules which consist of atoms in a variety of configurations. The atoms in turn contain a nucleus with electrons rotating about it, the nucleus itself being composed of many charged particles.

It seems as if scientists are determined to find the lowest common denominator of life, that sub-atomic particle or unit of energy or whatever, of which all matter consists. Our poem seeks to address the essential nature of this thing called matter.

In my discussion of matter I included energy, for matter and energy are closely related, as attested by Einstein's famous formula $E=Mc^2$. In fact the conversion of matter into energy was dramatically demonstrated in the atomic bomb. Recently scientist have also witnessed the conversion of energy to sub atomic particles, or matter. I also consider light and sound to be forms of energy and therefore included within the term matter.

But besides matter, the stuff of which all things are made, there exists on this earth the whole realm of consciousness and thought. These cannot be dissected or examined under a microscope as matter has been. Consciousness, the ability to think, and the whole realm of spirit is an enigma to the physical scientists. By and large they have kept these things outside the parameters of their investigations.

What is consciousness? Anything aware of its existence is conscious. So consciousness is awareness of existence. Thought requires consciousness. Awareness is thoughts of existence. Consciousness and thinking go together. You can't have one without the other. You can't think without being conscious. You can't be conscious without thinking about your consciousness. The whole realm of consciousness and thought I have called the spirit realm.

Relationship of Spirit and Matter

What is the relationship of spirit and matter? Is matter conscious? It seems to me that consciousness cannot exist in a vacuum. There must be something that is conscious of its existence. There are degrees of consciousness in matter. A human being consists of a complex arrangement of matter and is highly conscious. An animal is conscious but probably to a lesser degree than a human. Even plants seem to have some degree of consciousness, while a rock would appear to have little if any consciousness. Sub atomic particles in laboratory experiments appear to be influenced by the consciousness of the experimenter. There is a relationship between matter, consciousness and thought or between matter and the spirit realm.

Although God has been traditionally thought of as spirit, this poem says God is both matter and spirit and in a unique relationship. Man also is a spirit being but is intended to form a relationship with matter to become like God upon the earth. The poem suggests that matter is that part of both God and man that has been minimized or overlooked and is actually referred to as the "stone the builders rejected" who becomes the "chief cornerstone" (Math. 21:42).

The Source

The poem starts out by saying that all that exists

today did not come from nothing and therefore has always existed. It may not have existed in its present form. But the totality of all that exists has always existed. "There is nothing new under the sun" including the sun. The form may have changed, the arrangement may be different, but the totality has always been.

Scientists in their theory of the conservation of energy and mass have attested to the idea that matter never increases or decreases but has always existed in its totality. In the spirit realm the poem brings out the idea that all spiritual realities have always existed. All the good in the world and all the bad in the world have always existed. "Something cannot come from naught. So everything that is, both of matter and of spirit has always been." Consciousness was never invented it simply is.

The source is simply the starting point of our thinking. We have not considered the source of the source but understand that it also always existed.

The source of all that is must have contained within itself all that existed of the material realm and also of the spiritual realm. There has always been this relationship between these two realms.

Our poem talks about this relationship and calls this source "glorious thinking conscious matter, the source of all that is". "And from this thinking came all thought and from this knowing everything that ever could be known and from this being everything that ever there could be".

Polarities

Within the spirit realm there are elements that could be considered positive and others negative. Both must have been contained within the totality of the source. If good and bad exist they have always existed within the source. Love exists and has always existed. So all that is not love has always existed within the source.

Polarities exist such as light and dark, hot and cold, male and female, good and evil pleasure and pain. Because they exist they must have been contained within the totality of the source.

When polarities exist together they tend to

neutralize each other. The poem pictures the source in its totality and neutrality in an anthropomorphic manner as being "quite bored".

"It was everything but wanted to be something so set about to change and re arrange itself". Boredom and ambition to become something are characteristics we may well have inherited from our source.

The source spoke the Word and the Word was with God and was God. This whole poem is about God becoming. The process of God becoming is ongoing and is not finished yet, nor will it ever be.

The use of the term God and the pronoun "Him" is clarified in the introduction.

Evil does not exist except as a polarity of good. The separation of good and evil is necessary to know good. Evil will disappear when we join good and evil, this is what it means when it says "darkness could not overcome light". Experiencing good and evil on this earth is the way of joining good and evil and eliminating evil. So experience is the key. It is necessary for man on earth to walk through evil. This is why we must partake of the fruit of the tree of knowledge of good and evil. Partaking of this fruit was all part of the plan, not the "fall" of mankind.

As with good and evil so with all polarities, light and dark, happiness and sadness, peace and turmoil, death and sickness, freedom and bondage, plenty and lack, joy and sorrow, pleasure and pain, love and all that is not love.

We should not say that polarity does not exist, neither should we face polarity with fear and guilt, polarity existed in God (the mind of God) from the beginning. Mankind will overcome as God overcame.

The Mind of God

Everything is matter or spirit. The source contained both matter and spirit or matter in a relationship with spirit. The spirit or consciousness of the source could be called the mind of God. The mind of God is the spirit realm.

Like an artist sees the completed work of art in his mind before ever a stroke of the brush touches the canvas, so God saw it all in His mind first.

The whole of creation of the universe was seen in

the mind of God. But God knew that thinking would not make it so. Thus he spoke the Word to bring about the necessary separation of spirit and matter and the power of love resulting in the big bang and the separation and dispersal of matter to form the material universe so that what was in His mind became reality.

So also God created in his mind a vast array of entities of thought each with consciousness as He had and seeming almost as alive as He, yet all within the confines of His mind.

It says that we were thoughts in the mind of God long before we appeared in material form. In this sense we were predestined in the mind of God before the world began. It is my understanding that we will eventually conform to the image that was in His mind a long time ago, although our free will may delay this process.

The whole spirit realm exists and has always existed within the mind of God. When we die we return to the mind of God. When we incarnate we leave the mind of God and become like God upon the earth, spirits in relationship with matter and in a material universe.

The whole message of this poem is that while our

source is in the mind of God our destiny is in our relationship with matter as God was in His original singularity.

Separation

The story of all that exists is a story of God becoming. Our becoming who we are is part of His story.

For God to become something (although He was everything) there had to be a series of separations in the totality of what He was. Nothing could be added to God for He was everything. He could only become something by separation.

Our poem describes some of these separations. We speak of separations in the sequence of time although they may have taken place in eternity or not in the sequence we describe. Remember, this is an allegory and not necessarily a true description of events as they happened. Neither is this a complete description of all the separation that may have taken place or be taking place. The becoming of God is an on going process. We are not complete so God has not completed His process of becoming.

The whole creative process is a process of separation not building up something out of nothing. We create a painting by separating the color from the tube, separating a part of the landscape from the whole. Everything is there to start out with. The statue is within the stone. The excess stone simply needs to be chipped away by the sculptor's chisel. God Himself was everything and became something through the process of separation. His creative acts were cuts of separation. Everything was there to start with. To be created it simply had to be separated.

A few specific separations that occurred to make things the way they are, are mentioned in the poem as follows:

Separation of Light and Darkness

"And God separated the light from the darkness" (Gen. 1:4). "The darkness was not able to over come the light" (John 1:5). These passages of scripture are probably the most significant words ever written. It is not just a description of a step in the creation of the universe but a momentous occurrence in the process of God becoming. God did not create light out of nothing. He was light and darkness and created light by separating light from the darkness. You could say He became light by separating Himself from darkness. Neither did God create darkness. Light and darkness always were. Of course darkness is the absence of light, but the absence of light always existed. Darkness is the polarity of light.

It is hard to conceive of the mixture of light and darkness. Perhaps this would be as close to nothing as we could conceive. Can you see how important a step this was in the "becoming" of God? Nothing could be without this first separation.

The mixture of light and darkness was formless neutrality. This was probably a reasonable description of God before he decided to become something. Can you see why I was inspired to say in my anthropomorphic tale that God was probably "quite bored"? With this separation clarity came out of the blur of existence. Matter begins to take on form and shape, shades and shadows define the reality of substance. Space comes into being along with time.

Now for the first time we can see the dark side of God but also His radiant glory. It is most reassuring to know that "the darkness was not able to overcome the light". This is why we can know with certainty that everything is going to turn out all right in the end, however ominous it may be to know there is a dark side to God. A friend of mine was inspired to write a song called "Darkness is the Garment of the Lord", without this garment of darkness we would not be able to discern the shape of a God of total light. However bright God may be He would have no shape without the garment of darkness. Evil is a shadow giving good its shape.

Separation of Good and Evil

For thought to take on substance it must take on polarity. It must know good and evil, light and dark. Another separation in the process of God becoming was the separation of good and bad, or more succinctly the separation of love from all that was not love.

Our anthropomorphic God looked at this quality called love and decided that of all the things He would like to be it would be love more than anything else. But He knew that in order to be love He would have to separate love from all the residue that was not love. This residue was a hideous sight and no doubt contained all the woes and pains of all the ages and generations of all time.

But God knew the nature of reality, that love had the edge and would eventually triumph as light triumphs over darkness. So He decided this separation could do no harm. He, in His singularity would always choose love over the residue.

So the Word went forth again and love was separated and God became love and is to this day. This in effect is the explanation for the existence of evil. It may not be the whole story but for me it is a better explanation than any I have heard so far.

Separation of Spirit and Matter

Next we speak of the separation of matter and spirit. In the poem there follows a description of the creation of the universe in the thinking of God. God it seems explored the vast potential of His mind. It says He revelled in His boundless imagination and pictured galaxies and stars and worlds in space. The fantasies of His mind seem to know no end.

But it says He realizes that thinking alone would not bring these wondrous things into reality. Another separation of His totality would be required. I spoke earlier of all things consisting of matter and spirit So God who was everything consisted of substance and mind (matter and spirit). His mind could conjure up the most fantastic images of suns and moons and worlds and atoms, molecules etc. But it would require His substance to form such things in reality. This universe of His mind, if it were to go beyond His thoughts would have to be constructed of the substance of what He was; matter, First came the separation of matter from mind then matter (which was God as well as His mind) from itself.

We have often spoken of the creation of the universe as a miracle of God where God created it out of nothing. It was in fact a stupendous scientific accomplishment in which God, the nuclear engineering physicist supreme, created a universe out of Himself (His own substance).

Scientists have come up with a theory for the origin of the material universe which is widely accepted. They seem to agree that the universe had its beginning about 15 to 20 billion years ago with a big bang. The big bang they say was an atomic explosion surpassing imagination which set in motion the forces creating the universe. There was, they say a singularity of highly condensed matter which exploded releasing energy and scattering matter in a chain reaction resulting in the ever expanding universe that is observable to astronomers today.

Our writings say that this singularity of highly condensed matter was none other than the substance of God Himself. (Not the mind of God).

Nuclear physicists would never attempt to guess the amount of energy that would have been required to bring about an explosion of the magnitude of the big bang. Our poem simply says "the only source of such a vast amount of energy to do the job was love". God was love for He had already separated love from all that was not love. So it says there began a chain reaction precipitated by love to create the joyous explosion we call the big bang and the resulting formation of the universe.

The universe was formed out of the substance of God which separated from the mind of God and from itself. The singularity of God became a billion or more sub atomic particles, the building blocks of the universe. So what He created in His mind became real by separating the matter of God from the spirit of God. God thus gave Himself to bring this about.

This was the Lamb that was slain from the foundation of the world. (Rev. 13:8)

This was the inheritance of Abraham when it was spoken that his offspring would be as the stars of the sky and as the sand of the sea shore (Gen. 15:5). What else is as prolific as the stars and the sand than the sub atomic particles of the big bang. God's sacrifice of Himself to create the universe is pictured many times in scripture.

The particular one mentioned in the poem is Gen.15 where the sacrifice was divided and separated and the burning lamp passed between the pieces.

But the mind of God remains in-tact to this day. While the substance of God is manifest in the universe. But God is still "all and in all". "All" in the glory of His mind and "in all" the substance of His universe. (Eph. 1:23)

Separation of God's Thoughts From Himself

But God was alone in His glorious universe and anthropomorphically felt His solitude. You could say He was lonely. So He set about to create in His mind a being like Himself with whom He could share His thoughts. He became quite familiar with this fantasy of His mind who He called "Lucifer". He conferred upon this spirit of His mind ability to think as He and ponder and create. And between the two of them they thought up many more such beings in their mind. Eventually there was a cloud of souls each fashioned individually in His mind as an author would create in his mind many characters for his play. It says that you and I were also created in the mind of God before the foundation of the world. This is born out in scripture (Eph. 4:1 f.f.).

What follows is a conversation between God and Lucifer in which Lucifer acknowledges his wonderful place in the mind of God but points out that he is not free nor is he real, and openly wonders about this thing called "matter" and what treasures may be hidden therein. You could say Lucifer put a "bug" in the mind of God about making the creation of His mind to be as "real as He". God considered the proposition and concluded that it would indeed be a superior creation if endowed with His substance and free to be themselves, "Gods in their own right".

But the only catch was that having separated love from all that was not love. Would these living souls choose love and leave the residue as He did?

He knew of course that if they chose wrongly they could choose again and right would triumph in the end but in view of their freedom He could not tell how long. God being the adventurer that He was decided to take the chance and thus was launched the great and wondrous experiment of the ages. So God separated once more, "His thoughts from Himself, their thoughts from His and their will from God's will". So the vast creation of His mind became free spirits.

Now He must prepare a way for them to join themselves to matter to become as God in substance as well as spirit. So God prepared a place for His spirit creation to become like Himself. A planet called earth within a solar system of the milky way. A billion years of preparation and evolution until upright creatures walk upon the earth. Then comes the incarnation of the mind of God into bodies prepared for them. Described at some length in the poem.

The Separation of Male and Female

As has been pointed out God was everything so He contained within His totality both male and female qualities. You could say He was androgenous. We need to more often refer to God as "Him or Her" or "Mother-Father God". As God, so Adam was androgenous. God saw that it would be better for Adam to be separated and have a companion of the opposite polarity. Thus began the wonderfully superior arrangement of the separation of the sexes.

The Fall

These writings are in total contradiction to the traditional doctrine of the fall of man. The woes of the world are not the result of something man did in contravention to the will of God. Man did not disobey God by eating of the tree of knowledge of good and evil. On the contrary it was intended and necessary that man eat of this tree. It was necessary for man to look at all that was not love in order that he may choose love above all else to become like God and walk as God upon the earth. To be like God it was necessary to be free and freedom meant the ability to choose. God had in mind the highest and the best for man. To achieve this man had to choose for himself. God was well aware of the possibility of some choosing something other than total love. But they would have the chance to choose again and again until in the end (and God knew this) right will triumph over wrong.

In our story Adam hesitates to eat of this tree but Eve (as she did in the Genesis story) jumps right in and eats and gives to Adam. Then Adam and Eve, as did all who partook of the fruit of the tree of the knowledge of good and evil, saw it all. They saw as never before the wonder and the beauty of the love of God. But they also saw the "residue", the dark side of God. Adam reacted with fear, though Eve not quite so much. This was not totally unexpected by God. For He knew the end from the beginning and although Adam at first chose to fear, he would have the right to choose again and yet again. Which no doubt he did in the eight hundred years he lived upon the earth before he returned to the spirit world. When he came back to this earth as the second Adam (Jesus) He became the first one to perfectly demonstrate to all the world man's ability to choose love above all else.

You will notice in our poem that Lucifer (being of light) the first born in the mind of God becomes Adam, first to incarnate in human form. We also have Jesus the second Adam as a reincarnation of the first Adam.

Therefore none other than Lucifer. My apologies to tradition which has Lucifer cast out of Heaven to become Satan.

Satan is not the source of evil but is a spirit of God's mind with a special purpose to personify the "residue of all that was not love". Satan didn't get a part in my play. You'd think I could have worked him in somewhere!

Conspiracies

Many conspiracies arose in the course of time which deceived man and delayed their achieving the purpose of their incarnation. It started amongst those who followed Adam in partaking of the fruit of the tree of the knowledge of good and evil. The story says that those who ate saw even worse things than Adam in the "residue that was not love". They saw avarice and pride, selfishness and greed and deceit as well as fear. They yielded to the fear they felt. Thinking no doubt that they should prevent others from eating of this tree they yielded also to deceit. They thought to spread the word around that God had actually forbade them to eat of the tree of the knowledge of good and evil. Then they concocted the story that if men ate of the tree they would surely die, as if to die was bad. It was of course a lie.

But fear of death arose and spread although God meant for death to be a simple natural way to get back to the spirit world. Soon guilt and fear ran rampant through Paradise where God intended naught but love.

The conspiracy of guilt for knowing good and evil is widespread even to this day. But it is not a sin to look upon the residue of all that remains when love is not present. It was always God's intention for men to see evil so that he could wisely choose to do good.

The tree of the knowledge of good and evil had to be planted by God for man's benefit, to enable him to make informed choices. We ought not to feel guilty for having eaten of this tree or allowing ourselves to be exposed to evil.

Neither is it God's intention for man to feel guilty even for indulging in the "residue". For if we make a wrong choice we will have learned a lesson and will be able to make a better choice next time. We should not experience guilt but be challenged to do better.

Sin has been made to appear as if it were something deserving punishment or even death. Whereas it is simply missing the mark. A child in an archery lesson with his father as teacher would certainly not be punished or even scolded for missing the bull's eye. Rather would he be given loving instructions on how to aim and shoot so as to come closer next time.

Our Father would like us if we fail to try again in this lifetime and if needs be even in future lifetimes as often as is necessary until we learn the wisdom of love.

Conspiracy of Control

Some, when they looked at the "residue" didn't see the need for simple choice. It was something to be regulated they thought. We will make laws and rules, and regulations for men to obey. Of course along with rules came penalties for disobeying the rules and courts to judge and jails to punish.

Out of all these arose the conspiracy of control. Men became their own worst enemy. They took it upon themselves to guide and control their brothers. Many thinking they knew better sought to impose their will upon others. They took it upon themselves to condemn those they felt had done wrong and all the while overlooking their own mistakes and covering up their own guilt.

Some took control and tried to drive men from the garden of Eden, "lest you eat of the tree of life and live forever". Look upon your sweat, and toil and fears of snakes and sickness and dying most of all as punishment from God. Paradise upon the earth is not for you they said.

This conspiracy has been most effective. For hardly any have eaten of this blessed tree of immortality. Yet it was the intention of God for man to live forever on this earth or perhaps to come and go at will. It was not the flaming sword of an angry God, but fear and guilt that has kept men from partaking of the tree of life to live forever.

But worst of all there arose the conspiracy to imitate the voice of God. They did this first with Adam and Eve in the garden. "Adam where art thou?", they heard while walking in the cool of day. It was not God who spoke to make them guilty of their nakedness and run and hide though they had done no wrong. We'll try it on the rest they thought and so arose a company of those who took it on themselves to speak for God.

They are with us to this day. Zealous to control and direct their brothers they enjoy and abuse the authority of God. So widespread is this phenomena that men have forgotten they can hear the voice of God themselves. To hear God speak is suspect and subject to their interpretation of the "Book" they wield with skill and prejudice.

The Conspiracy of the Gods

There is strong evidence to suggest that there have been powerful entities from other realms who have come to this planet and have set themselves up as gods. Hierarchies of such beings we have called God and given our obedience and worship. It has been suggested that they have tampered with our DNA through biogenetic rearrangement. Which has resulted in the curtailing of man's abilities on earth. In this manner they have been able to control man to suit their own agenda.

There is also strong evidence to suggest that their rule is coming to an end. Electromagnetic energies from Prime Creator through cosmic rays are entering the earth's atmosphere. This results in a restructuring and restoration of our DNA, and a re-order of man's capabilities on earth.

I am not qualified to speak of all of this nor have I any revelation on the topic. However I do believe it to be true and feel that genetic restoration can only help in the process of spirit relationship to matter and their enlightenment and empowerment as Gods upon the earth.

The Conspiracy of Angels

There were those of the spirit world who came with Adam to spy out the land to see if indeed spirits of the mind of God can take on form and walk in freedom as Gods upon the earth. Some returned with a bad report. The giants are too strong to overcome. Would it not be better, they said, to encourage spirits to stay where they are and even tried to get those who are in the world to return to spirit realms. Although misguided many are well meaning and sincere. This conspiracy has continued even to this day and is more widespread than you think. They have spread the word skilfully in celestial realms as well as on the earth, that Heaven is the true home of spirits and this is where they belong.

First they have tried to devalue matter. Far from seeing matter as particles of God with which spirits can have a lasting relationship for which they were destined, they portray matter as evil. They say only spiritual things have any real value. They do not see the body, for instance as a temple of God but rather something that needs to be put down. "The flesh" has become a synonym for evil.

They say the body is a shell housing the spirit which alone is of real worth.

Volumes have been written promoting this belief and much even has been included within the sacred scriptures of most religions. The conspiracy is as old as man upon the earth. This conspiracy has described our stay on earth as temporary. Nor can they offer a satisfactory explanation for why we are here. Earth is a school house some say for us to learn lessons then return to spirit realms. But what lesson could we learn here to help us in that place where all is bliss from whence we came. Some who have returned to the spirit realm and have elected to stay speak even from the other side their messages of celestial sedition.

Some portray life as an illusion a dream from which we will awake in due course. They say we are aliens here, our Father's house is our home and the world is not real.

But everything is right here and now, not in Heaven after death. Death is leaving home not going home. Birth is coming home to try again. This is it. This is Heaven. Let's wake up and see it. We are spirits of God in a body of God, in a world of God, in a universe of God. Conscious matter is the stone the builders rejected, the lamb that was slain from the foundation of the world. God cast Himself out of Heaven, the realm of His thoughts, and is waiting patiently for his thoughts to return. Matter is the cause, thoughts are the effect. The spirit realm is part of God but separated from the Father's House, this material this long long journey? Could it be our eyes will gaze upon a new earth as Adam gazed in wonder the first day, a paradise restored, beautiful in form, beneath a panoply of love to cover all?